You Matter in this Information Universe

Rev 3

Which may Contain No Matter Particles; Just 'In-Formed' Energy

JESUIS LAPLUME

Ordering Information:

Prime Seven Media
518 Landmann St.
Tomah City, WI 54660

Printed in the United States of America

Our Milky Way is a galaxy
and our planet earth is on one of its arms,
but is fully connected
not only to all of this galaxy
but all others as well.
We are a One in The One,
but few understand how connected
that we really are!

Table of Contents

Why You Should Read This Book

You are being taught a whole lot of very outdated ideas about what this universe is like, as well as your special place within it as a human.

That makes it very difficult for you to live a Great life and become a very Great humans being.

While this mini-Book presents a very different, leading-edge viewpoint, it is presented to you so that you can reconsider both **Who** you are and **Why** you are here.

Not enough is provided, on purpose, for you to use this mini-Book as a roadmap, but just to challenge almost everything which you may presently believe; because you have been so badly informed by society.

A great deal of work on how our universe has formed, as well as how the smallest items of what it is made of, suggest that our previous understandings about both us, and our universe, are turning us **away** from Truths; **not towards** them.

Why I Wrote It

As a writer guided to investigate the latest in Science and Philosophy, and then write about what I have discovered or have been asked to write, I still get called to add to my Books already written. Although I have been working on other Books and mini-Books, I was asked to both develop this one and to then publicize it.

A General Caution

As one who has studied many fields,
and learned about the limits in all of them,
please take the following caution to heart:

**It is both your right *and* your responsibility
to accept from the things that anyone offers you,
only those that will make you a better you.**
*What those will be is between you and your Maker
– no one else.*

Please take all that the author writes
as a reason to form new **questions** about **Who** you are,
as well as **Whom** you are meant to be.

*There will only ever be ONE you
and the universe will not be as Great as it could be
until you become as Great as you can be!*
This is the universal Truth of being human.

Please Note:
You are a complex being in a complex universe.

Please read my books through once – quickly.
I write in a style that forces you to expand your horizons
by then reading slowly, carefully considering the
caveats presented to you for your consideration.
Absorbing a wider perspective is necessary
for you to fully become both Love and Compassion.
Adults can do that; frightened children won't!

The Story to Be Told

I am not asking you to accept that these ideas are **true**; but only ask what you might change in your life if they were **true** and you are now willing to try new ideas.

You have been taught that we are matter beings in a matter universe, based upon old, outdated science and parts of much older philosophy.

At the leading edges of science and philosophy, some of us are cluing in upon the more likely reality that there are no matter particles, as such, and that the universe is primarily a **Pure Information** one. It guides a relatively small subset of what we call energy, to act either like fields, or as what we think matter particles should act like. It is almost certain that there are no real matter particles; but rather a subset of organized energy!

Long before we were able to look so deeply into our universe, in both space and time, we formulated equations which helped explain how that universe would behave. We also developed instruments which could look at the results of processes at very small scales and develop theories and equations that organized what we could say based upon such ideas.

Now that we have much better instruments to study our universe at the universe and atomic levels, our findings, as well as our older theories and equations are being refuted. This disturbs some older experts and really excites some of the younger ones.

I am presenting my own, **personal viewpoints** about both the creation of the 'stuff' all around us and what is really going on at the universe at galactic scales. Although they may be accurate, I present them as my present stories about both our universe and our **special** place as humans within it.

I am NOT asking you to **believe** or **accept** them, but rather to just open your mind, heart, and soul, to the possibility that such a perspective might help you lead a better life!

In summary, this may well be a universe with no matter particles within it; it is our minds and senses telling us useful stories about what is going on, at all scales.

One of our ongoing issues in understanding all that really exists is that we presently lack adequate words and definitions for the non-physical sixth senses which we have; as well as the five or more physical senses which science has taught us so much about.

There are probably no dimension words or theoretical constructs for a thought, an emotion, an intuition, a belief, or any of the processes we collectively lump into the spiritual processes.

Some very recent science is building evidence of seemingly instantaneous connections between things like photons of light and some bits of energy we call matter; because we presently lack better words for them.

With all this going on, evidence is building on many levels that our old ideas and predictive equations are no longer holding up; but the inertia of the scientific bureaucracy is slowing down the brilliant young Science types who are finding ways to do great work anyway.

The young are trying to lead the old; but the old refuse to be led! This process is as old as humanity itself; but may be set aside as soon as new discoveries explode into being; this in spite of the barriers and fortifications still being built and reinforced by those present 'experts'.

So, in a way that this old man can, I will lead you part way through the process of what **may** be real, and may actually be going on. Hopefully you will find it both exciting and yet comforting; if I do a good enough job.

To lighten things a bit, please ponder upon an earlier "Jesuisism" of mine:

"If you aren't laughing at life, you don't understand it well enough – Yet"

The Yet is there because there is always **hope**! (especially for a "Quarter Newfie" (i.e. someone with one grandparent from Newfoundland - like me!)). LoL!

1

What Existed Before The Big Bang Event?

❁

Our founding assumptions about what the universe is really like require that we investigate the **Context** within which everything happens; but science types usually run for cover when asked to do so; although some suggest that philosophers would be more likely to hazard a guess. Over time, and from many sources, many who consider such things have suggested that all we 'see' around us, even out and back to the beginnings of the universe, would appear to be a greater intelligence than that available to any human-related being.

When I wrote my Books "Alone?: No One Can Be Separated From God" and "You Are In God: God Is In You" I called upon a God which was not anthropomorphic (like any human). One of the suggestions is that the universe was within God; because there was no 'space-time' place, before the process which 'looks like The Big Bang' came into being.

More than a decade later, and after a lot of thinking and about it, and pondering upon it, I take nothing back about the concepts in those books. I have learned a lot since then, but nothing refutes there being something before that event and it was, - to all intents and purposes - unlimited in any way, shape, or form. This, you will find if you ponder upon it, is at least a little bit scary.

There must have been a field of Pure Information which was powerful and extensive enough to manage the very, very rapid expansion of 'whatever'. That was needed so that structures could start to come into being in an orderly-enough way. They would still appear to be evolving almost 14 billion years later.

Such a God, by the way, need have none of the limitations or characteristics which so rightly many reject as being too-limited; even for a human. We humans do like to invent limitations on everything; largely so we will not feel too helpless and threatened. The God I am suggesting would have none of the nasty characteristics that some humans still invent (or refer to) to justify their nastiness! I will expand upon that later; but please try to consider a God of Love who would have us be more than just stardust. We humans do have bodies made of stardust; but we are not limited to that state!

Please reread this short book, **several** times, to determine where my presentations are less than effective in that there is a lot to cover in this mini-Book!

It is both your right and your responsibility to be skeptical about what I have presented!

2

Information Has No Dimensions From Science

✾

many who write about spirituality choose to reuse words for spirituality and sixth-sense processes. They do this because there are no universally accepted words for some of the concepts. This invariably results in confusion that is not helped much by that misuse.

What is the dimension of a thought, an emotion, an intuition, a belief, etc.? We can do brain scans and now even locate where the brain is active when such things occur; but we cannot access, nor properly name, any of those processes. We do use our brain to do some local processing of what we are thinking about, or feeling, intuiting or acting upon. We know where such things are activating blood flows and see the production of electric or magnetic fields when they are happening. But we cannot pull out the thought or feeling itself, let alone save it to elsewhere.

One of the great strengths of science, as it evolves over time, is our properly defining the processes going on within our body, especially those related to our five physical senses. We can even prove that some dogs can sense a single molecule of some important smells/odours in a dog's life (therefore proving how inadequate we are at such tasks).

You can infer that another loves you, often in many ways, but nothing from science can pick those 'bits' out and save them to another place. Romance writers have largely expanded what various love-related processes are. They can be partially communicated; sometimes over huge distances between the parties experiencing them. Some of the processes may be entanglement, but are called chemistry; but no chemical molecule has been definitively identified and stored for later study.

Given the interest in such things, perhaps breakthroughs will occur and trigger developments in other sixth-sense processes. Love is a powerful drive in humans; even some kinds of love are very important in other species. Knowing how much we **don't** know can be a great driver for better research.

I was shocked to find that there is much work needed to even list the components of our sixth senses; guys may not be as interested but gals surely are!

I would have attempted to write this mini-Book much earlier if I could have found more out there about our non-dimensional senses. I have experienced many of them myself, but sure would like company. Perhaps you readers could help assemble a related database and send it my way to jesusislaplume@gmail.com LoL!

3

A Universe Which Selected For Life

hen we look at the equations which humans have developed to describe our universe, we find at least 24 parameters involved, some of which had to be very accurately developed to have the set work as well as it does. When all has been said and done, the evolution of life, if only random chance were involved, makes any life existing in this present 14 billion year old universe appear to be impossible.

However, it certainly appears that life did evolve billions of year ago. If life exists against what appear to be virtually impossible odds, one should not be discredited for asking whether pre-existing fields of Pure, non-dimensional Information included a built-in set of instructions to favour the early development of life. Any other position seems irrational and an intentional rejection of the existence of an agency, or set of instructions, which made the evolution of Life inevitable.

For childish reasons, after the worst of the Spanish Inquisition, a set of science types declared that no such fields were allowed to exist. For a while that was a direct declaration. Later it became a more insidious inference. Inferences are both more difficult to identify and also more difficult to effectively dispute.

We are still, presently, in that situation. Many science types act as though they will not accept the existence of any top-down fields that would guide which bottom-up processes are allowed; or even preferred! I stand with those who say this is largely a fear-based and irrational stance to take. Given what we now 'know' well enough, what is highly unlikely, at both the universe and sub-atomic levels, is such a set of processes presenting themselves to those with open minds.

Unfortunately changing the closed minds of many, older and entrenched Science and Philosophy humans takes time; or perhaps just their very-early retirement.

Many of the younger science types seem quite open to things outright denied by their superiors within science-related organizations. The young assume that Philosophy does have a place in guiding Science; but the established top dogs often deny that this can be so!

As a physical system Science type of person, by both training and practice, I long ago found that what older

Science and Engineering folks insisted **was so** was often ridiculously off base. Established 'experts' often do not like to have older Science or Engineering challenged; they take that as a personal affront. Truly-great ones act differently, because they know, and fully accept, the reality of how little that they know.

'Know-it-all-types' don't!

4

The Universe
Then Selected For
Special Beings

here is a huge range in the complexity of living creatures, all the way from a bacterium to the most evolved humans who have ever lived.

David R. Hawkins, PhD, invented/discovered a log based 10 'Scale of Consciousness' to apply values to how evolved 'things and life' were. That scale for life on this beautiful, blue planet goes from 0 to 1000. Other species of life forms may have evolved to go well past 1000!

As we learn more about other species of life on our planet, we are getting results which make our human minds much less superior than those of some other primates, birds and such as dolphins,, etc. However, our human ability to develop and ask questions, such as the W5H ones (Who,

What, When, Where, Why, and How) do seem to be largely present **only** within human beings.

In my opinion, the two most critical ones for our leaving Fear behind and moving up to Love are two of the 'W' words in the W5H set:

"**Who** are we?" and

"**Why** are we here?"

It is vital that we, who wish to evolve out of primal **fear**: a) ask both those questions; b) learn a lot about the issues involved; and c) develop our own, personal answers to those questions. After that stage, we may well gain more by asking the other, four W5H questions.

Each human being provides a somewhat different, actually unique, viewpoint on what it is like to be a living human being. As one who has studied the variability that all of the Galaxies provide, the universe seems to be 'into' both variability and complexity, as opposed to identical copies. Some humans seem stuck on the concept of clones of 'perfected' beings; but this universe does not seem to treasure that approach.

When we approach what is knowable with fully open minds, it is likely that there are other, **special** species in this universe. For reasons which I will bring up later, they may already be among us; waiting for us to evolve to the point where meaningful interaction with us is possible. **Fear**-based humans are not a useful component of humanity for such an interaction.

The sub-set of humans who are presently attempting to live **love**-focused lives may include a further subgroup that would be able to interact with any such 'aliens' in a useful way.

Becoming a great example of a **love**-focused human could have amazing consequences!

I invite you to act upon becoming such a special human being soon! Time may be of the essence!

5

Our Sixth Sense Abilities/Skills

O ur bodies have five or more senses which provide us with inputs about both our environments and, usually, whether or not that environment is safe. Those senses include: sight; smell; taste; touch; and hearing. **Sciences** have been developed for each of those five senses and dimensions assigned to them.

We have other senses and skills, which are variously listed and cannot be measured directly. These sixth senses include, as I list them in their amount of power: Thoughts; Feelings; Intuitions; Remote Viewing/Sensing; Consciousness; and a broad and complex array of abilities which I bundle under Spirituality. The Thought ability is the one easiest to teach; so most humans know more about it than the others.

The sixth senses have no dimensions assigned by Science, so we have adjectives, like weak, strong or very powerful;

but **no** actual dimensions. We cannot measure a thought; even though we can use brain scans to determine where they elicit blood flow changes; both where in the brain and how strongly. However, we cannot isolate, measure or record the actual emotion, idea or belief. We can certainly talk about how we react to them, but they have no dimensions like the five senses have' scientifically-defined, measurable dimensions!

In his book, entitled "Real Magic" Dean Radin, PhD, listed several PSI processes/experiences: telepathy; clairvoyance; precognition; and, psychokinesis.

He has done the research needed to show that at least some humans have such abilities. He has proven that they are not electromagnetic; because they occur, unabated, when the source, receptor, or even both, are located in properly-shielded rooms. Each of these processes, plus others, are 'Pure, non-dimensional Information' processes. Recent work by others have strongly suggested that they are instantaneously found remotely. They are everywhere and do not become weaker over distances. The five senses and electromagnetic fields usually drop off strongly with distances (often by a cube-law rate). The strength of Information processes do not!

Based upon Dean's work, telepathy (which is sharing of images/emotions over separated minds) is real; as is clairvoyance (which is remote perception of events and images).

Precognition (which is the before-hand perception of distant events/images in time) is also real; as is psychokinesis (the influence of mental intentions on objects). Young and not-yet-damaged minds can deal with such concepts; but not many established, elderly, Science-only, types can, nor will they, do so.

6

Implications of The Double-Slit Experiments

✳

When light is shone against a set of slits which are close together, light goes through both slits and recombines on the other side. The waves interfere with each other, to make a set of peaks and troughs which we call interference patterns. This is a well-known process and we can calculate a lot about the patterns, based upon the distance between the slits and the length of the waveform(s) or frequencies of the light being beamed at the slits.

However, if we observe the processes, that interference effect does **not** take place, with the light gathering as though the light were now particles. This change is called the observer effect; but it also occurs even if the observer is just physical equipment designed to mechanically 'observe' the ongoing process. The observer effect occurs **even** if the 'observer' is just a **machine**. It took humans a while to notice this phenomenon, but it should have had us more-seriously reconsider just what, **who** and **why** an inanimate device can cause an observer effect.

Sometimes light behaves exactly how we expect waves to behave; but other times light behaves as if it were a stream of particles. If we send the waves towards the slits, one waveform at a time, the resultant patterns occur according to whether or not the process is being `observed/measured (by either a human `observer, or an inanimate device). It occurs as if the name of the effect should have been a recorder effect.

If we humans come to grips with what is going on when a process is being recorded by a device, or a human, it would appear that the universe itself In-Forms the process that the light can/should behave differently when what is approaching the slits is being, or is **not** being, observed/ recorded.

A long time ago, some philosophy types told us that everything in this universe is connected in very basic ways. Traditional Science does not yet take that connection seriously; but usually even denies that it can be so. Surely, they say, light cannot `know' whether or not it is being observed/recorded, especially one individual wave-form at a time! How could it possibly be so?

Simplistic, bottom-up analysis does not support such effects. Traditional Science was modified, after the Spanish Inquisition, to deny even the possibility of the existence of any top-down organizing fields. That was an emotional response to the Inquisitors, done in Fear that there could be something like a deity that the inquisitors 'said' supported their behaviour (although they were just trying to re-establish control of thought, by any others than priests!). Oops!

7

Energy Which Looks Like Matter

❋

As we learn more and more about our universe, especially it's earlier times, it is becoming obvious that there **may** be no actual matter particles in the universe. There may be just some **energy** which has frozen in form (In-Formed, as it were) and it behaves like what we though real matter would look like (so we call it matter-like).

When you knock your knuckles against a table, no matter particles are there to cause the 'feel' of skin-over-bone on the wood of the table top. The fields that behave like electrons powerfully and forcefully reject each from others getting too close to each other; our brains tell us a story about our knuckles smarting and the table top being hard. It is a useful story; but not an accurate one.

As the universe evolves, new arrangements of sets of fields form and change, but **no** matter particles are involved.

Some **energy** fields are In-Formed and represent what is actually going on. The sets of fields are just that; the stories are very useful and even exist when other living creatures experience interactions. That does not make the stories correct, or precisely accurate; but they are both useful and persistent over extended periods of time. One of my Books elaborates on that:

https://www.amazon.com/Species-Who-Tells-Stories-Everything/dp/1544801424/ref=

8

Our Minds as Information Processors

✻

We live within a universe which is primarily Information, with a part that is energy (plus In-Formed energy) which behaves like we thought matter should do!

In this universe we have minds, (gifts from the universe and not products of our brains), which are efficient and effective 'Information processors'. But we have not yet been effectively taught this way of perceiving the universe and our human minds within it.

Our human minds are much more than random outputs of **energy** fields and make us **special** in an amazing universe of Pure Information (something that has no dimensions from the **energy**/Science domains).

There is no dimension for: a real thought; or emotion; or intuition; or for consciousness; or for the very real, but misunderstood, processes from Love and Above that **spirituality** presents to us.

Not having dimensions has many who are interested in **spirituality** miss-use names that are legitimate in Science; but are not appropriate in any process beyond what Science can describe. This causes great confusion! The confusion is especially troublesome when those who miss-use the words don't yet fully describe what they mean by those miss-used terms.

There is NO 'frequency' for a process in intuition, nor in consciousness, nor even in the **spiritual** domain of 'Love and Above' processes. Incompletely evolved minds are NOT effective or efficient at dealing with this; however, that is **where we are**, at this moment in time.

The 'Scale of Consciousness,' developed by David R. Hawkins, PhD., can determine the power of a statement about all things; it can be a great tool once you learn how to use it. If you simply believe that there can be **truths** in this universe, you can learn to use that 'Scale of Consciousness' technique to discover some **truths**.

So, for a relook at our universe, plus our **special** place as humans within it, for the moment please consider the following:

- This is an Information universe:
- Our minds are gifts from the universe;

- They are efficient and effective in processing the Information which is present within the universe;
- Most of what you have been taught, or developed on your own, is flawed, however; and,
- We can choose to understand both **who** we are and **why** we are here; then build upon that to grow further in the comprehension of 'All That Is' (a way of describing the universe)!

9

Our Connection to
The Cosmic Mind

o that others can control you, you are falsely taught that we are **alone** and vulnerable to those with power and vast riches. The reality is that everything in the universe is connected and we are part of that vast **One**.

Not only is there a component of the universe which acts as a **cosmic mind**, but our gift from the universe, our eternal **mind**, is always directly connected to that **cosmic mind**. We have been effectively taught differently, so that others can control us; with billions spent each year to ensure that we are surrounded with false teaching which keeps us vulnerable to those with power and riches.

Minds are **not** energy items or processes, but 'Information processes' and are not as limited as energy (and the special forms of energy we call wrongly call matter). Information does **not** drop off with distance, but is everywhere, un-degraded, and the connections appear to occur instantly (the more precise our measurements, the more 'instantaneous' the connection appears to be). This process is called Entanglement and recent, well-done research in China is showing it to be very real!

Rewording our mind connections very slightly; our human **minds** are fully, continuously, connected to a '**cosmic mind**' which is everywhere in the universe. Through our human **minds**, therefore **we** are instantly connected to everything in the universe.

Other **special** species, if they exist, would also be instantly connected to everything - everywhere.

A little bit of contemplation will help us understand why those who try to control us, using **fear**-based mechanisms, would not want humans to ever understand that this could be (or is) **true**; nor what that means for each and every one of us. It is likely **true** that this is a real description of what is going on. Those, however, who recognize that they are eternal, **spiritual** beings, who have bodies in a physical form, are not going to be controllable using **fear**-based methodologies.

If humans who have power and vast riches got them by unethical means, surely they will not want us to hear, certainly **not believe**, that we are beings who are connected to 'All That Is' and now 'know' it. No one can control a human or other species who fully understands such a reality!

We humans, who 'know' who we really are, will be uncontrollable; so we are 'seen' as dangerous to those who need to control everyone, and even everything, to 'feel' safer.

You be the best you can be and let them deal with their own fear-based issues!

10

Our Minds Guiding Development of Our Brains

❋

Our human **minds**, our gift from the universe, pre-exist our physical bodies; and can sometimes be controlling/using our brains within those bodies.

If you choose to think about it, for even a short period of time, and accept that **minds** are everywhere, surely there are evolutionary advantages to having our human **minds** guiding the development of the physical brain within our bodies. If you build a house, and can guide the design and construction of that house, it is almost certainly true that the house will better suit you as soon as you 'move in' as it were.

Those who 'believe' that brains don't create **minds**, can likely come up with a better understanding of child prodigies who can play piano like an expert; as soon as the can span the keys with their small hands. The stories of humans who can remember past lives, will then become much more understandable, if we have many lives; so that we can finally get our journeys '**Right**'!

Many societies, from supposedly-primitive to very-advanced ones, have 'creation' stories which include our coming back in different bodies; each one appearing to have a more mature understanding of what a life can, and should, become. Just because we don't **yet** know **how** that can be possible does NOT reject any such possibility!

11

We Are a One in The One

✽

Because of how the whole universe came into being, with everything connected, although moving apart at significant speeds, we are all a **One** in **The One**.

No separation is possible. Any story which says differently is just nonsense from the ignorant ones!

However, there are usually **many** ignorant ones!

Those ignorant ones 'Know it all' in their own, tiny, minute minds and they treasure being minute and powerless (so they can blame others for their issues); but we are actually neither! However, their belief that they are powerless allows them to act as they don't have the ability to become a **great One** within **The One**.

The universe may have **no** limits; but we humans can *falsely invent* limits; to excuse our lack of growing towards

Becoming Love Personified. Excuses are just ignorance, although now-verbalized!

Even great prose and convincing arguments do not make falsehoods into real **truths.**

You are here within **The One** and likely have no limits. Think that through and reconsider just how far you could go; if you just got up and started onwards!

Inventing limits does not make you great; but does make you too frightened to 'see' how far you could go if you just started to evolve '**Right Now**'!

The Story Retold

We humans are one of a set of **special** species who can question everything and evolve to the point where we can even change physical reality. We are a **social** species which has become very powerful by learning to cooperate in ever-larger groups, but we often allow a few *fear*-based individuals (and sets of individuals) misdirect us away from becoming All We Can Be.

At this very special point in time, we can continue to be ruled by the powerful and vastly rich. These **fear** that 'they are not enough, just as they are', and so choose to compete with other humans, as well as against the processes of life itself to get 'ever more'! We humans could choose to reject that false teaching, then choose to become a **Love**-focused species that treasures fully living in **Love**-focused ways instead.

In competition with other humans and life processes, we will destroy our ability to grow in the power of **Love** and become **great**; or we can succumb to the nonsense of competition with all that makes life great, and so helping *the few* destroy the ability of all humans to survive the negative impact of those *few*.

Given the failure of most of our education systems, at so many levels, to teach us all about the reality of our ignorance, humanity can only survive if a significant subset of us step up and fight for a more widespread understanding of how complex this universe really is; as well as how little we 'really know for sure about it' and also our '**special** place within it'.

Many humans have chosen to try to fully-live **Love**-focused lives and that subset of humans (about 6% of all living humans) is already more powerful than all other humans combined; including the powerful and very rich. However, that set of 6% is not yet well organized.

This universe can only be properly modelled using non-linear tools like logarithms, base 10 (the norm we have used so far).

Using such tools, we can predict how a few percent of all humans combined can be more powerful than all other living humans. This is not only unsettling to the powerful and vastly rich, but also to almost all of humans who have not yet been properly taught about their **special** status in this universe of ours.

Each and every one of us who learns how to step out of the weak ways of *fear*, and start their journey to becoming **Love**-focused, can significantly add to the power of the existing subset of **Love**-focused humans to change how humanity influences our ultimate fate as humans. We can also change the fate of our life on this beautiful, blue planet which we call Earth.

Or we can *do nothing* and *lose* all that matters!

What to Do About It

In spite of the general reality that most humans have already been taught to just sit back and let others rule our lives, as well as what we are doing to the planet and life on it, we humans are far from being impotent! We are very **powerful**; if we choose to act in **Love**-focused ways. What we cannot afford to do is to remain *ignorant* about what is going on, or our ability to change processes in amazing ways!

Learning More About Information

This is not really an **energy**/matter universe, as much as it is actually one where 'Pure, non-dimensional Information' In-Forms a portion of **energy**. We have the real power behind all that is going on. Our **minds**, given to us by the universe itself, are Information processors which can steer how some of the **energy** is In-Formed in ways which can quickly and powerfully change what damage we humans have already done.

To learn about Information, we will have to go back and dismiss as inadequate almost all that we have been taught about the reality of physical matter being what really 'matters'; given that there may actually be no real matter particles in this universe of ours. It is only some **energy** that is In-Formed to act in ways we thought matter would behave.

That is a scary process, but a wee bit of insight into the many sixth-sense abilities we can have and use will be sufficient in getting us started.

Verifying What You Know About Us

Although we are complex beings in a very complex universe, the **truth** about humans is that we are one of several **special** species who can choose to leave *fear* behind and become **Love** focused; by rejecting virtually all virtual limitations taught to us.

In an Information universe, where our **minds** are Information processors, we can choose to perceive ourselves as capable of growing into the **special spiritual** beings we are destined to become. You can personally choose to leave all false imitations behind; by a simple choice to do so. Societies will try to change you back into a *fear*-based version of a human; but you can simply '**Keep on Keeping on**' being a more evolved version of whom you and we are and what we can become.

Teaching What You Know

It is in helping others ask better questions about **Who** we are, and by becoming a great example of living **Love**, that you can teach others who they are. In the end, many humans accept what they're told; but we can all choose to mimic those who present us with an example of living **Love** and is and also therefore be ones who refuse to accept false teaching. You are both within the universe and infused with it, including the portions of the **cosmic mind** with which we all can interact.

There is *no other*; we are all part of **The One!**

Walking The Walk

It is only in becoming great examples of living **Love** that others will find us worth mimicking; thus continuing the revolution in growing the size of the component of humanity who are using our **minds** (and their connection to the **cosmic mind**) to perceive reality; not accepting the errors and limitations *falsely* taught to us so that others can *control* us for their ends, not ours.

It is not only humanity that needs us to become as **great** as we can be; but all who live on this beautiful blue planet!

Start Your Journey into Becoming Love Now!

The Author

Y ou should always know something about anyone who proposes to teach, or show, you something of value. Something of me follows.

I was trained as a Scientist, but in an Engineering faculty – this sort of training is best suited for those who might then like to help transfer new Science into new Engineering practice, or help move new answers into more common practice. It is short on specifics and long on basics, including the need to question everything, even (or especially) the questions themselves. My basic Bachelor of Applied Science (Aeronautical Option) degree was used to its fullest in eight different fields, but was of most use was in aircraft and spacecraft component design. In the spacecraft area I performed not only a review of the systems design of the Canadarm on the US Space Shuttle but helped with the detail design of the six joints of that complex electromechanical mechanism.

When working in the field of housing and health I also learned a great deal about our ongoing limitations in present social and societal processes, not to do what was 'right', but rather a propensity to choose what was self-serving and anti-life.

Here the system science questions were more widely-based; the errors of our ways were more devastating for both mankind and the beautiful blue planet that gives us the opportunity of a great life.

Because I am often writing in areas that are well beyond those that I have been trained to know, from traditional sources, I would like to ask the reader to take to heart the cautions in **A General Caution** at the beginning of this book.

What The Author Believes

As the result of decades of study in both science, as well as the spirituality part of Religion that defines **who** we are and **why** we are here, etc., I have formed some understanding of the big questions. I have also determined that we **know** little but **believe** a lot. It is important that you understand where any author is coming from, so I will present a short list here, of what I believe, so that you are forewarned as to my inherent biases, if they are not obvious in my actual writing,

I believe that:

We are spiritual beings having a human experience, so that we can work through the conflicts that will result in our growing greatly in spiritual power.

However, the bureaucracy of religions has effectively perverted the spirituality that was the reason for their religion forming in the first place, so that they can now control their congregations; through **fear**.

The basic spiritual messages common to most religions are essentially correct, but we have to weed out the distortions introduced by the bureaucracies of our present religions.

We can never die, but our bodies will die – we are eternal spiritual beings because the Creator has chosen to make us so – don't tell God what She/He can't do!

Traditional science has turned some of its original assumptions into Dogma that can no longer even be questioned, so that science needs to be massively reconstituted to become relevant and honest again (science should never contain **any** Dogma that cannot be questioned).

The internet and new communication systems will allow an explosive growth of an understanding of our true nature and purpose.

Big industry and government is destroying the ability of our planet to support our existence and we must become truly loving creatures; or we will consign ourselves to the status of a failed species; one which failed opportunities to become great.

Actually the list could go on and on, but that should give you fair warning about where I am coming from.

If I were to meet you I would try to Love you just as you are. This is likely the highest goal that most of us will achieve in this lifetime, even though more is possible. Only 0.4% of human beings now living (still a whopping 28 million in 2010 when this was determined) are capable of unconditional love, and use that as their way of living. I am not there yet!

I am not a recognized teacher or guru, just a student and *a possible guide for other students.*

God gave us free will so that our choice to be great would not be fated and trivial, but wonderful and powerful. Be a lover, an unconditional lover, and study your real reasons for being. Then go ahead into the uncomfortable regions with your fear under control. Fear passes away when you go right through it to **truth**.

Remember, only the foolish and stupid have no fear. Only the brave face their fear and head on out anyway. Be brave and then be proud. That sort of pride is born of humility and is part of it. It is the only type of pride of self that is acceptable.

References & Other Sites

Since I am attempting to establish a leading edge viewpoint on the origins of our universe, as well as our own **special** place as humans within it, I do not have references as to others who are attempting to present similar resources. What I do have, however, is several sources which you could use to connect to the consciousness of the universe and the Truths thereof.

My primary resource is the late David R. Hawkins, PhD, and whose widow keeps his products available on his website at:

www.veritaspub.com or

https://veritaspub.com/

If the Ctrl + Click does not work on your server, please copy and paste one of the above to get to His site.

You do *have to believe* that there are **real truths** available in this universe for his methods to work.

A secondary source of access to the consciousness of the universe is Eckhart Tolle and he has a few **Free** courses (he also has more extensive, paid ones) available at:

https://beingthelight.eckharttolle.com/awaken_your_inner_light_free_mini_series42533397?_ke=

Again, if you don't believe that the universe can be conscious, the course and process may not work for you.

The Author's Sites
And Books

The Author's New Web Site

www.jesuislaplumecdnisbn.com

Author's Page on Amazon

www.amazon.com/Jesuis-Laplume/e/B00DCFMZD0

(The 0s in the above address are numbers, not letters)

(Or search for 'Jesuis Laplume' on www.amazon.com)

The Author's Larger Books
(By Order Published)

"On Life And Love And Why We Are Here

An Introduction To the New Spirituality

For Beginners"

"The Affirmation Prayer

Affirming Who You Really Are!"

"Alone?

No One Can be Separated from God"

"You Are In God:

God Is In You"

"Does Science Present The Truth?

"Truth" Vs. "Good Enough"

"The World Is Not How We 'See' It

It's Really Quite Weird"

"Learning To Be Great

An Introduction"

Four Brains, Four Egos And Many Minds:

Understanding How We Think"

"Successful Change:

In A Universe That Is Change"

"The Species Who Tells Stories:

And Everything is Stories"

"How Love Can Vanquish War:
  Love Will Win The Day"

"Loving Yourself First:
  Is This The Way To Go?"

"Unconditional Love:
  A Primer"

"Changing The Universe:
  By Changing Yourself"

"You Are A Powerful, Eternal, Spiritual Being:
  You Are One With The One"

"Hate The Sin:
  But Not The Sinner"

"Becoming A Courageous Adult
  Or Staying A Frightened Child"

The Author's Mini-Books

"Science And Spirituality:
  No Conflict At All"

"Who We Really Are:
  And Why You Are Here"

"You Were Destined To Be Great:
  Don't Be Stuck At Very Good"

"Living in Love:
  Or Just Surviving In Fear"

"To Love

And Be Loved"

"Your 'God'

is Too Small"

"You Matter

in This Information Universe"

"Who are we Humans

and Why are we Here?"

Afterword Postscript

Hopefully what you have read was useful and has had you reconsider new possibilities about your Special place in this amazing, but poorly understood and described, universe

Like all non-fiction books, it will be better and more useful to you after you choose to **re-read** it and **ponder upon** the messages provided within it. *Repetition* and *note-taking* are powerful tools when it comes to developing new understandings!

You may learn more by discussing this book with others, but carefully choose those people; most humans who live Fear-based lives can be seriously hurt by having their understanding of **Who** they are and **Why** they are here questioned. Only share the book with the 'open to new things' people that you know.

If you change, because you have read the book
several times, and pondered upon it, others may
ask you about why you seem different these days.

We can best get others to grow by
having them choose to mimic us,
rather than by telling them things.

Teach by becoming a good example
of an evolving human being!
Everything in this universe was
given a form of Free Will.

Trying to control others always goes against that
reality of Free Will; this is a Cosmic Truth!

If you leave reviews, positive or negative,
please take the time to partly explain why your
review is as it is. You can help others if you
clearly edit your review and keep it honest!

The world needs more Honesty!

www.ingramcontent.com/pod-product-compliance
Lightning Source LLC
Chambersburg PA
CBHW032103020426
42335CB00011B/475